POCKET ATLAS OF MRI
BODY ANATOMY

Pocket Atlas of MRI Body Anatomy

Thomas H. Berquist, M.D.
Associate Professor, Mayo Medical School
Consultant in Diagnostic Radiology
Mayo Clinic
Rochester, Minnesota

Richard L. Ehman, M.D.
Assistant Professor, Mayo Medical School
Senior Associate Consultant in Diagnostic Radiology
Mayo Clinic
Rochester, Minnesota

Gerald R. May, M.D.
Assistant Professor, Mayo Medical School
Consultant in Diagnostic Radiology
Mayo Clinic
Rochester, Minnesota

Raven Press 🦅 New York

Raven Press, 1185 Avenue of the Americas, New York, New York 10036

Made in the United States of America

The material contained in this volume was submitted as previously unpublished material, except in the instances in which credit has been given to the source from which some of the illustrative material was derived.

Great care has been taken to maintain the accuracy of the information contained in the volume. However, Raven Press cannot be held responsible for errors or for any consequences arising from the use of the information contained herein.

9 8 7 6 5 4 3 2

Library of Congress Cataloging-in-Publication Data

Berquist, Thomas H. (Thomas Henry), 1945–
    Pocket atlas of MRI body anatomy.

    Bibliography:
    Includes index.
    1. Anatomy, Human—Atlases. 2. Magnetic resonance imaging—Atlases. I. Ehman, Richard L. II. May, Gerald R. III. Title. [DNLM: 1. Abdomen—radiography—atlases. 2. Anatomy—atlases. 3. Extremities—radiography—atlases. 4. Nuclear Magnetic Resonance—atlases. 5. Pelvis—radiography—atlases. 5. Thoracic Radiography—atlases.
QS 17 B532p]
QM25.B47 1987   611   86-29649
ISBN 0-88167-282-3

Preface

This text is designed as a portable reference for anatomy in body magnetic resonance (MR) imaging. Images are displayed in the axial, coronal, and sagittal planes. It is especially important for users of MR to refamiliarize themselves with the coronal and sagittal anatomy as it applies to routine MR practice. Special emphasis is placed on the extremities where spatial resolution, coronal and sagittal planes, and soft tissue contrast provide more anatomic detail.

Most images in this text were obtained using partial saturation sequences ($T_E 25$, $T_R 600$). Slice thickness varies from 1.0 cm in the abdomen and chest to 0.3 cm in surface coil images of the extremities. Special parameters and positioning are noted below images when indicated. The anatomy is labeled using numbers with legends at the top of each page. An illustration demonstrating the level and plane of the MR images is provided with each MR image.

This volume will be of interest to physicians who are interpreting MR images or referring patients for MR studies and to fellows, residents, or medical students.

Contents

Upper Extremity 1
   Shoulder and humerus, 1
   Elbow and forearm, 14
   Hand and wrist, 22

Chest 28

Abdomen 43

Pelvis 56

Lower Extremity 69
   Thigh, 69
   Knee, 75
   Calf, 83
   Ankle, 88

Bibliography 97

POCKET ATLAS OF MRI
BODY ANATOMY

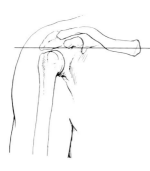

Body Coil
1. trachea
2. thyroid (left lobe)
3. sternocleidomastoid
4. anterior scalene
5. coracoid
6. deltoid
7. humeral head
8. subscapularis
9. cervical portion of spinal cord
10. trapezius
11. scapula
12. clavical and subclavius muscle

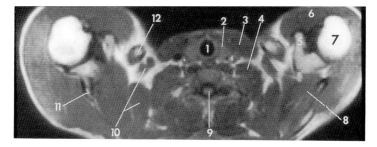

Surface Coil
1, sternocleidomastoid
2, deltoid
3, humeral head
4, scapular spine
5, supraspinatus

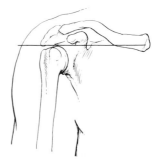

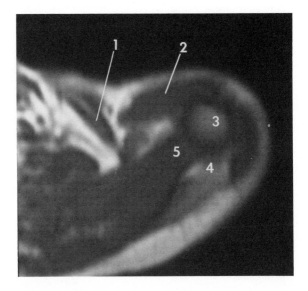

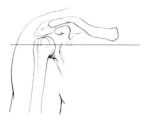

**Surface Coil (note
increased signal
posteriorly due to
position of coil)**
1, sternocleidomastoid
2, pectoralis minor
3, bicipital groove
4, deltoid
5, anterior glenoid labrum
6, posterior glenoid labrum
7, infraspinatus
8, subscapularis

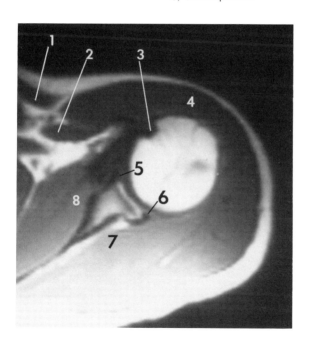

Body Coil
1, trapezius
2, supraspinatus
3, infraspinatus
4, scapula
5, subscapularis
6, axillary artery and vein
7, pectoral muscles
8, subclavius
9, clavicle

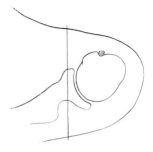

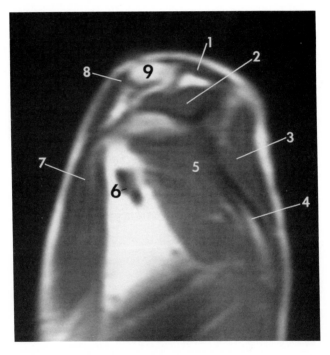

Body Coil
1, splenius capitus
2, trapezius
3, supraspinatus
4, scapula
5, infraspinatus
6, subscapularis
7, descending aorta
8, serratus anterior
9, teres major
10, deltoid

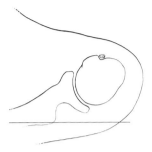

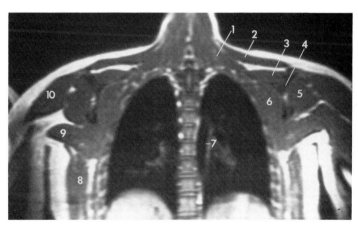

Body Coil
1, splenius capitis
2, trapezius
3, supraspinatus
4, deltoid
5, coracobrachialis
6, subscapularis
7, aorta
8, left pulmonary artery
9, teres major

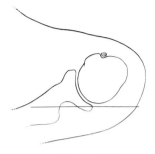

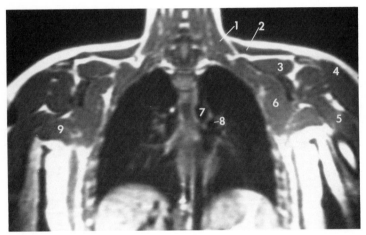

Body Coil
1, acromioclavicular joint
2, deltoid
3, aorta
4, innominate artery
5, innominate vein
6, axillary vein and artery
7, triceps brachii
8, brachialis

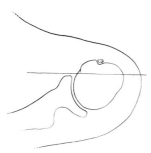

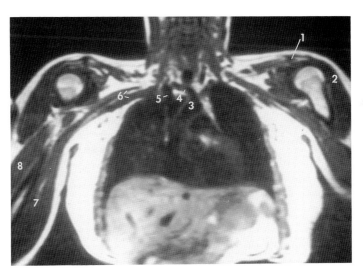

Surface Coil
1, clavicle
2, acromion
3, rotator cuff
4, deltoid
5, coracobrachialis
6, coracoid

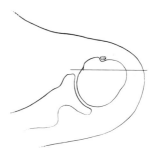

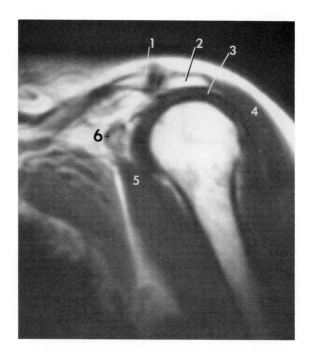

Surface Coil
1, pectoralis major
2, axillary vessels
3, teres major
4, humerus
5, deltoid

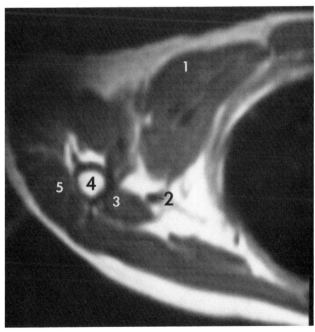

Surface Coil

1, biceps brachii
2, coracobrachialis
3, teres major
4, triceps brachii
5, deltoid

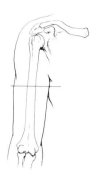

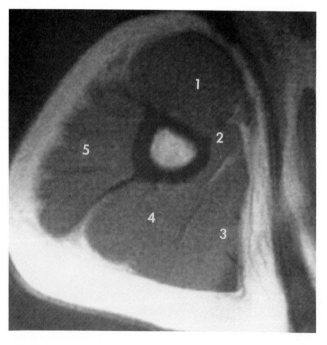

Surface Coil
1, biceps brachii
2, triceps brachii (medial head)
3, triceps brachii (long head)
4, triceps brachii (lateral head)
5, brachialis
6, cephalic vein

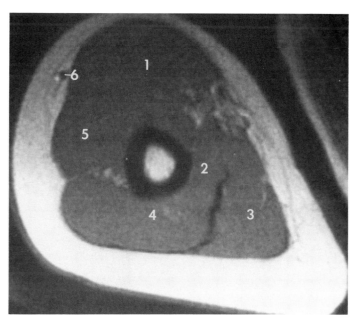

**Surface Coil (localized medial
view of mid humerus)**
1, biceps brachii
2, humerus
3, radial nerve
4, triceps (lateral head)
5, triceps (long head)
6, triceps (medial head)
7, ulnar nerve
8, basilic vein
9, median nerve
10, brachial artery and vein

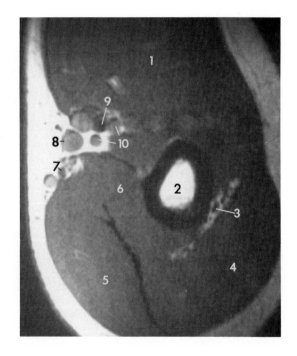

Surface Coil

1, biceps brachii
2, brachialis
3, cephalic vein
4, brachioradialis
5, humerus
6, triceps brachii and
   tendon
7, basilic vein
8, median nerve

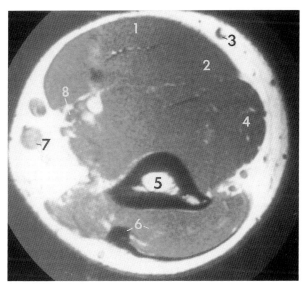

Surface Coil

1, biceps brachii
2, cephalic vein
3, brachialis
4, brachioradialis
5, radial nerve
6, extensor carpi radialis longus
7, dorsal antibrachial cutaneous nerve
8, lateral epicondyle
9, triceps
10, ulnar nerve
11, medial epicondyle
12, pronator teres
13, basilic vein
14, median nerve
15, brachial artery and vein

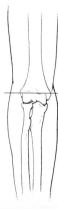

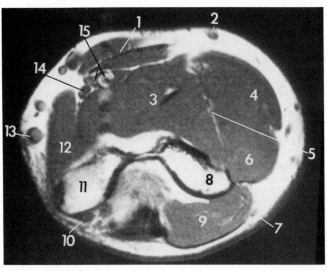

Surface Coil
1, brachioradialis
2, capitellum
3, anconeus
4, olecranon of ulna
5, flexor carpi ulnaris
6, ulnar nerve region
7, trochlea
8, pronator teres
9, brachialis
10, bronchial vessels

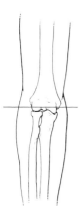

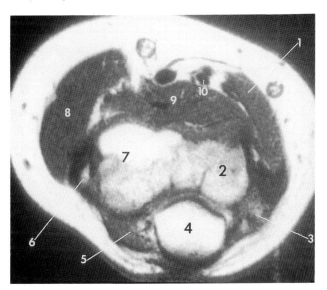

Surface Coil

1, brachioradialis
2, extensor carpi radialis longus and brevis
3, annular ligament and common extensor tendon
4, radial head
5, anconeus
6, ulna
7, flexor digitorum profundus
8, flexor carpi ulnaris
9, ulnar nerve
10, flexor digitorum superficialis
11, flexor carpi radialis and pronator teres
12, brachialis

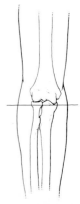

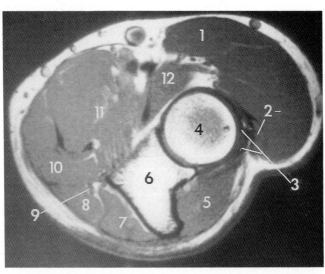

**Surface Coil (elbow
flexed and supinated)**
1, biceps tendon
2, radial tuberosity
3, ulna
4, radial head
5, capitellum
6, brachialis

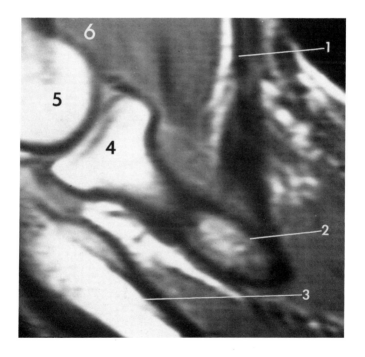

Surface Coil
1, trochlea
2, coronoid of ulna
3, radial head
4, capitellum
5, pronator teres
6, flexor carpi ulnaris
7, flexor digitorum profundus
8, supinator
9, brachioradialis

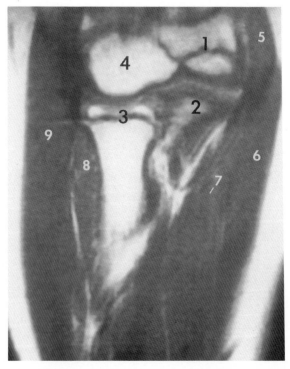

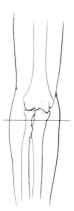

Surface Coil
1, brachioradialis
2, cephalic vein
3, extensor carpi radialis longus and brevis
4, extensor digitorum communis
5, supinator
6, radius
7, extensor carpi ulnaris
8, anconeus
9, ulna
10, flexor digitorum profundus
11, flexor carpi ulnaris
12, flexor digitorum superficialis
13, pronator teres
14, median nerve
15, biceps brachii tendon

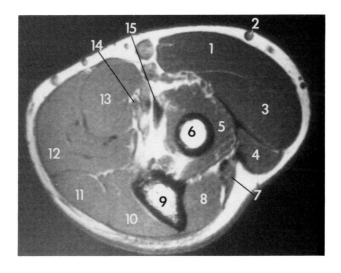

Surface Coil

1, flexor carpi radialis
2, brachioradialis
3, extensor carpi radialis longus and brevis
4, extensor digitorum communis
5, radius
6, abductor pollicis longus
7, interosseous membrane
8, extensor pollicis longus
9, extensor carpi ulnaris
10, ulna
11, flexor digitorum profundus
12, flexor digitorum sublimis

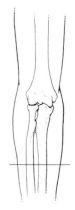

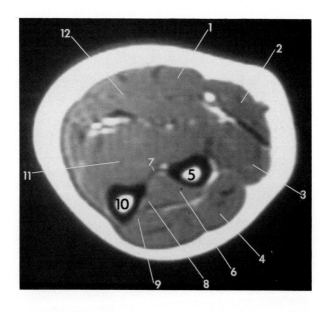

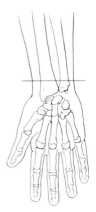

Surface Coil
1, flexor digitorum superficialis
2, median nerve
3, flexor carpi radialis tendon
4, radial artery
5, flexor pollicis longus
6, brachioradialis tendon
7, abductor pollicis longus tendon
8, extensor carpi radialis longus and brevis tendons
9, radius
10, extensor pollicis longus tendon
11, extensor indicis proprius
12, extensor carpi ulnaris
13, ulna
14, flexor carpi ulnaris
15, pronator quadratus

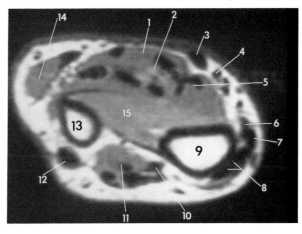

Surface Coil
1, flexor carpi ulnaris
2, ulnar styloid
3, extensor carpi ulnaris tendon
4, extensor digiti quinti proprius
5, extensor digitorum communis tendons
6, dorsal radial tubercle
7, extensor carpi radialis brevis tendon
8, extensor carpi radialis longus tendon
9, extensor pollicis brevis tendon
10, abductor pollicis longus tendon
11, radial artery
12, flexor carpi radialis
13, flexor pollicis longus
14, flexor digitorum superficialis and profundus tendons

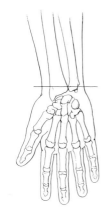

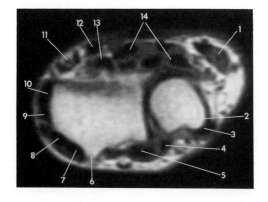

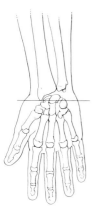

Surface Coil

1, flexor carpi radialis tendon
2, abductor pollicis longus tendon
3, extensor pollicis brevis tendon
4, scaphoid
5, extensor carpi radialis longus tendon
6, extensor pollicis longus and carpi radialis brevis tendons
7, extensor digitorum communis and indicis proprius tendons
8, lunate
9, triquetrum
10, extensor carpi ulnaris tendon
11, flexor carpi ulnaris tendon
12, flexor digitorum profundus and sublimis tendons

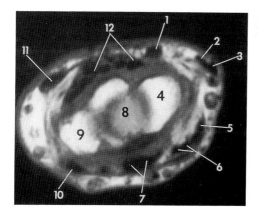

Surface Coil

1, trapezium
2, trapezoid
3, capitate
4, hook of lunate
5, abductor digiti minimi
6, flexor digitorum sublimis and profundus tendons
7, median nerve
8, transverse carpal ligament
9, flexor pollicis longus tendon
10, abductor pollicis brevis

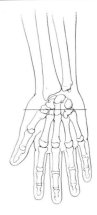

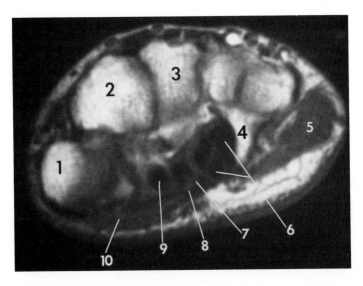

Surface Coil

1, extensor digitorum communis
2, 5th metacarpal
3, flexor digiti quinti brevis and abductor digiti quinti
4, opponens digiti quinti
5, superficial and deep flexor tendons
6, flexor pollicis longis tendon
7, adductor pollicis
8, abductor pollicis brevis
9, opponens pollicis
10, dorsal interosseous muscle

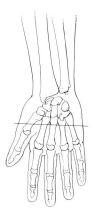

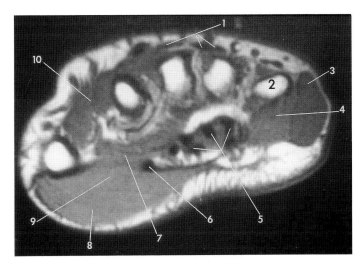

Surface Coil

1, flexor digitorum profundus and superficialis tendon
2, adductor pollicis
3, 1st metacarpal
4, trapezium
5, scaphoid
6, radial styloid
7, pisiform
8, hook of the hamate
9, abductor digiti quinti
10, opponens digiti quinti

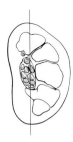

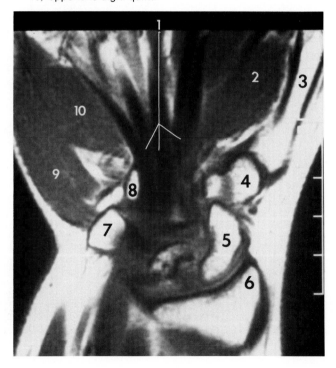

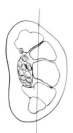

Surface Coil
1, trapezoid
2, capitate
3, hamate
4, triquetrum
5, ulnar styloid
6, distal radio-ulnar joint
7, pronator quadratus
8, radial styloid

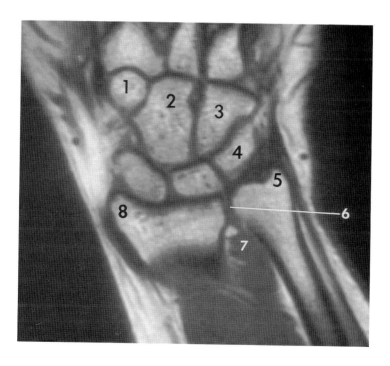

Body Coil (0.15 T)
1, left innominate vein
2, brachiocephalic artery
3, left common carotid artery
4, left subclavian artery
5, trapezius
6, trachea
7, infraspinatus
8, subscapularis
9, right brachiocephalic (innominate) vein
10, pectoralis minor
11, pectoralis major

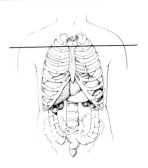

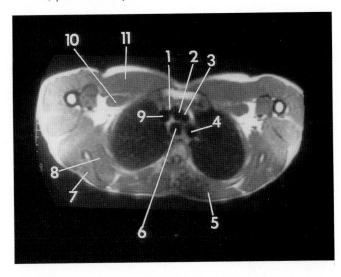

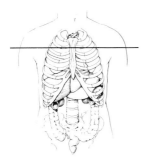

Body Coil (0.15 T)
1, pectoralis major
2, pectoralis minor
3, aorto-pulmonary window region
4, descending aorta
5, scapula
6, azygos vein
7, carina
8, subscapularis
9, infraspinatus
10, superior vena cava
11, ascending aorta

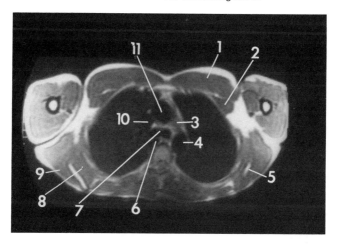

Body Coil (0.15 T)
1, pectoralis major
2, main pulmonary artery
3, left pulmonary artery
4, left main bronchus
5, descending aorta
6, azygos vein
7, right main bronchus
8, right pulmonary artery
9, superior vena cava
10, ascending aorta

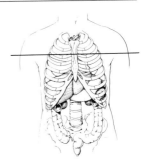

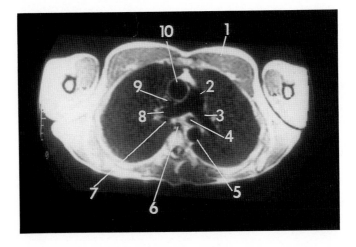

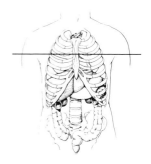

Body Coil (0.15 T)
1, right ventricular outflow tract
2, anterior descending coronary artery
3, circumflex coronary artery
4, left pulmonary vein
5, descending aorta
6, azygos vein
7, right pulmonary vein
8, superior vena cava
9, right atrial appendage
10, aortic root
11, pericardium

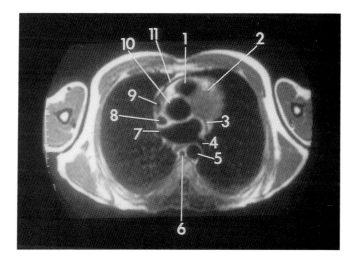

Body Coil (0.15 T)

1, ventricular septum
2, papillary muscle and left ventricle
3, mitral valve
4, coronary sinus
5, descending aorta
6, azygos vein
7, left atrium
8, right atrium
9, tricuspid valve
10, right coronary artery
11, right ventricle

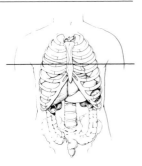

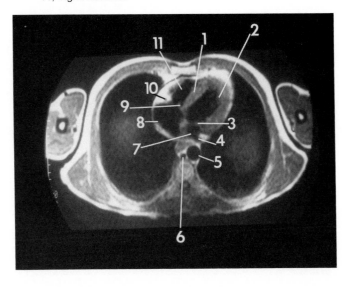

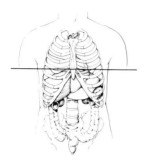

Body Coil (0.15 T)

1, pectoralis major
2, right ventricle
3, floor of left ventricle
4, coronary sinus
5, descending aorta
6, serratus anterior
7, azygos vein
8, latissimus dorsi
9, inferior vena cava
10, liver
11, right atrium

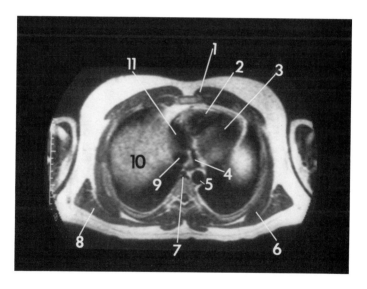

Body Coil (0.15 T)
1, azygos vein
2, intercostal artery
3, descending aorta
4, stomach

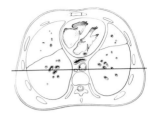

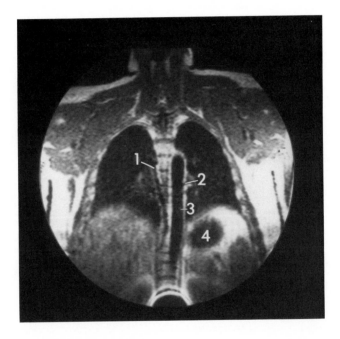

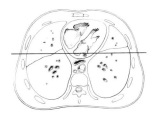

Body Coil (0.15 T)
1, spinal canal
2, aortic arch
3, left pulmonary artery
4, left pulmonary vein
5, left atrium
6, stomach
7, inferior vena cava
8, liver
9, right pulmonary artery
10, left main bronchus
11, azygos arch

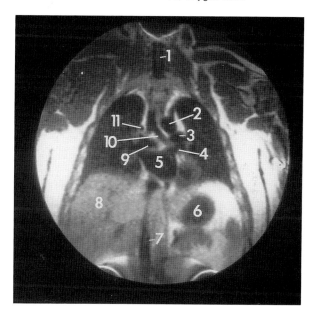

Body Coil (0.15 T)

1, right vertebral artery
2, left common carotid artery
3, left subclavian artery
4, aortic arch
5, pulmonary artery
6, right pulmonary artery
7, left ventricle
8, stomach
9, portal vein
10, hepatic vein
11, right atrium
12, left atrium
13, superior vena cava
14, trachea

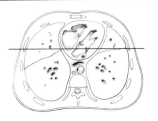

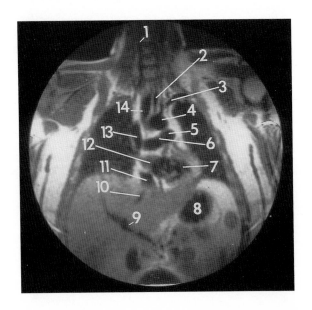

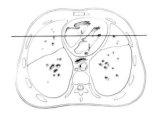

Body Coil (0.15 T)

1, larynx
2, trachea
3, left internal jugular vein
4, humeral head
5, left subclavian artery
6, innominate vein
7, main pulmonary artery
8, left ventricle
9, stomach
10, liver
11, right atrium
12, ascending aorta
13, right brachiocephalic vein
14, right subclavian artery
15, right internal jugular vein
16, right common carotid artery

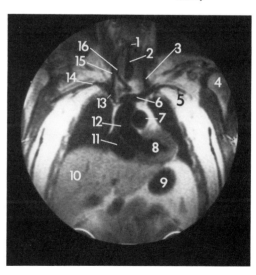

Body Coil (0.15 T)
1, pectoralis major
2, deltoid
3, left ventricle
4, right ventricle
5, stomach
6, liver
7, right lung

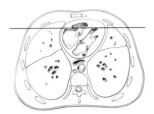

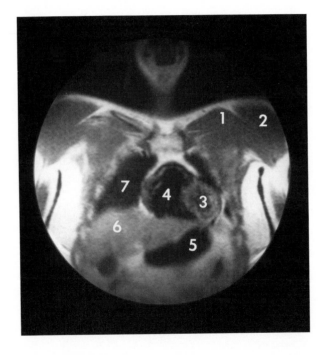

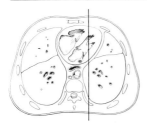

Body Coil (0.15 T)
1, sternocleidomastoid
2, left subclavian artery
3, left subclavian vein
4, left superior pulmonary vein
5, left ventricle
6, right ventricle
7, stomach
8, trapezius

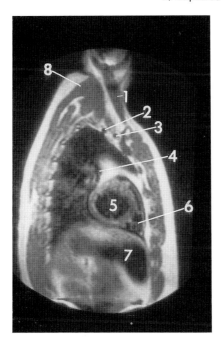

Body Coil (0.15 T)
1, innominate vein
2, main pulmonary artery
3, right ventricle
4, left ventricle
5, liver
6, descending aorta
7, left atrium
8, left pulmonary vein
9, left main stem bronchus

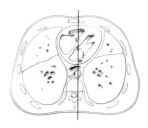

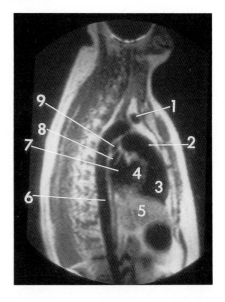

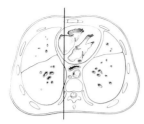

Body Coil (0.15 T)
1, trachea
2, brachiocephalic artery
3, aorta
4, right ventricle
5, right atrium
6, left atrium
7, right pulmonary artery
8, right mainstem bronchus

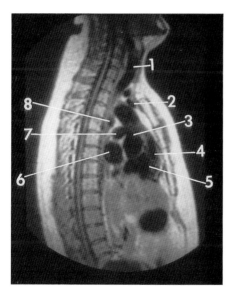

Body Coil (0.15 T)

1, right subclavian artery
2, superior vena cava
3, aorta
4, right atrium
5, right ventricle
6, hepatic vein
7, inferior vena cava
8, left atrium and right
   pulmonary veins
9, right pulmonary artery
10, right mainstem bronchus

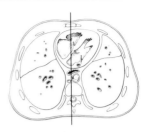

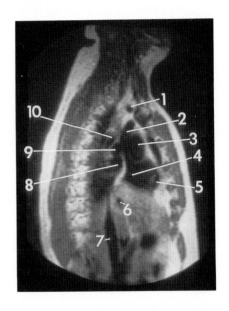

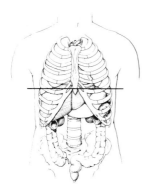

Body Coil
1, aortic flow artifact
2, heart–lower wall
3, esophagus
4, aorta
5, spleen
6, latissimus dorsi
7, longissimus dorsi
8, posterior right lobe of liver
9, serratus anterior
10, right hepatic vein
11, inferior vena cava

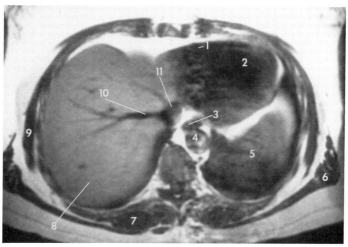

Body Coil

1, aortic flow artifact
2, left lobe of liver
3, stomach
4, spleen
5, latissimus dorsi
6, crus of diaphragm
7, posterior right lobe of liver
8, inferior vena cava
9, caudate lobe of liver

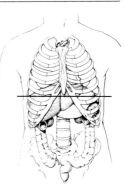

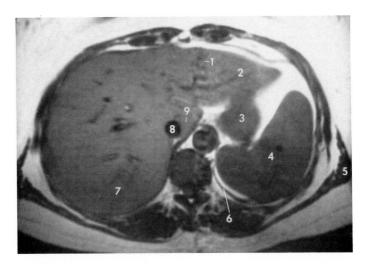

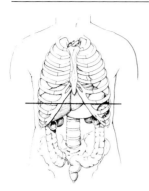

Body Coil
1. stomach
2. pancreas
3. splenic vein
4. spleen
5. left kidney
6. iliocostalis
7. longissimus dorsi
8. spinalis and semispinalis dorsi
9. right lobe of liver
10. inferior vena cava
11. portal vein
12. common bile duct
13. left renal vein

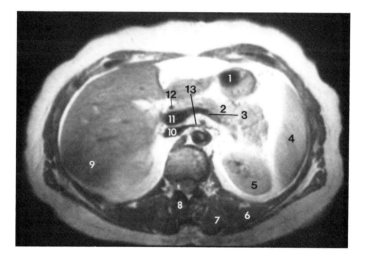

Surface Coil (adrenal level)

1, aorta
2, left adrenal
3, renal artery branch
4, crus of diaphragm
5, adrenal artery branch
6, right adrenal
7, inferior vena cava

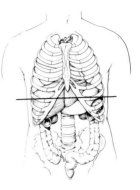

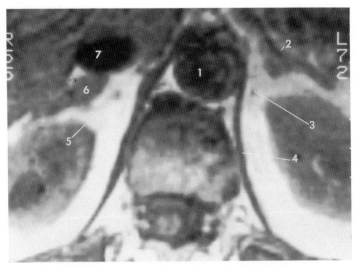

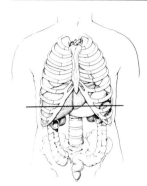

Body Coil (peripancreatic vascular anatomy)

1, stomach
2, pancreas
3, hemiazygous vein
4, azygous vein
5, inferior vena cava
6, portal vein
7, hepatic artery
8, celiac artery
9, splenic artery

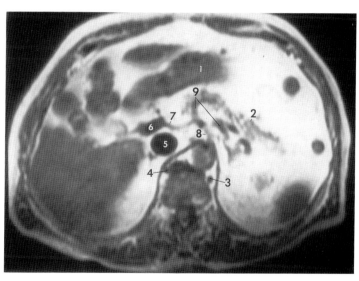

Body Coil

1, rectus abdominis
2, stomach
3, small bowel
4, left kidney
5, iliocostalis
6, longissimus dorsi
7, spinalis and semispinalis dorsi
8, psoas muscle
9, right lobe of liver
10, gallbladder
11, duodenum
12, head of pancreas
13, superior mesenteric artery and vein

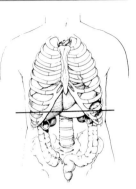

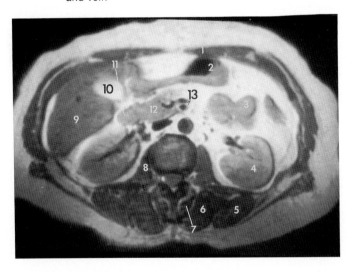

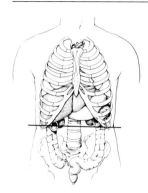

Body Coil
1, rectus abdominis
2, internal oblique muscle
3, external oblique muscle
4, latissimus dorsi muscle
5, psoas major muscle
6, lower pole right kidney
7, small bowel
8, inferior vena cava
9, superior mesenteric
   artery and vein
10, common iliac arteries

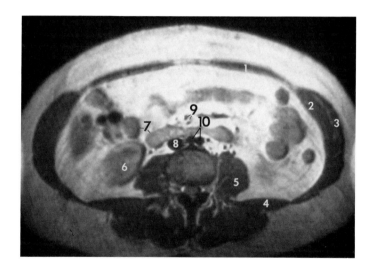

Body Coil
1, spleen
2, left kidney
3, splenic flexure of colon
4, psoas muscle
5, spinal canal
6, erector spinae muscle
7, spinous process L$_4$
8, gluteus medius
9, right sacroiliac joint
10, right lobe of liver

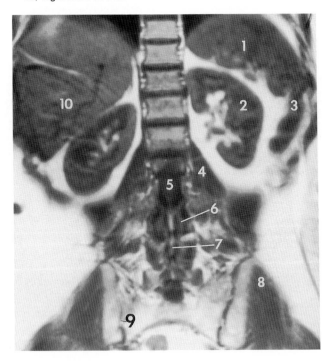

Body Coil
1, descending aorta
2, crus left diaphragm
3, stomach
4, spleen
5, kidney
6, psoas muscle
7, L$_4$ vertebra
8, external oblique muscle
9, liver

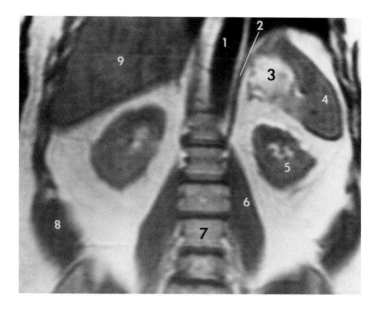

Body Coil

1, stomach
2, spleen
3, celiac artery
4, superior mesenteric artery
5, left renal artery
6, left common iliac artery
   and vein
7, inferior vena cava
8, liver

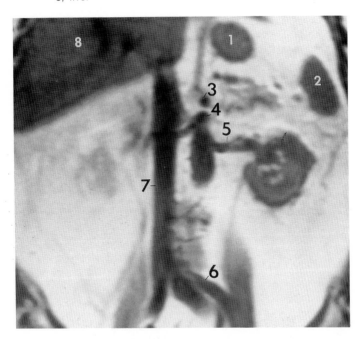

Body Coil
1, stomach
2, splenic artery and vein
3, pancreas
4, small bowel
5, psoas
6, pectus abdominis
7, transverse colon
8, diaphragm

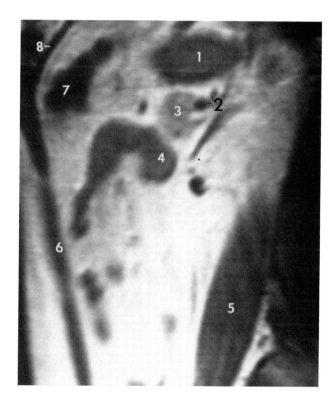

Body Coil
1, right lobe of liver
2, inferior vena cava
3, caudate lobe of liver
4, duodenum
5, stomach
6, diaphragm

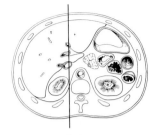

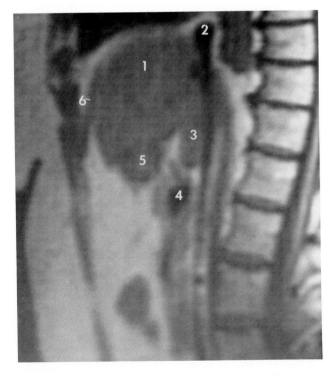

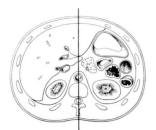

Body Coil (0.15 T)
1, abdominal aorta
2, hepatic vein
3, celiac artery
4, superior mesenteric
   artery

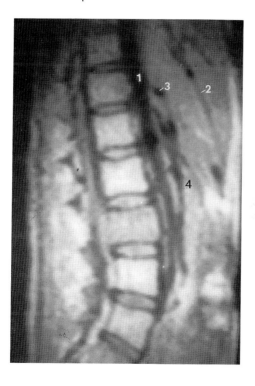

Body Coil

1, rectus abdominis
2, anterior superior iliac spine
3, gluteus medius
4, gluteus maximus
5, multifidus and iliocostalis lumborum
6, sacroiliac joint
7, femoral nerve
8, iliopsoas
9, gluteus minimus

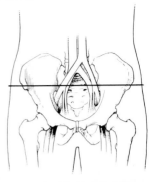

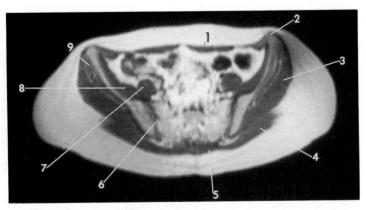

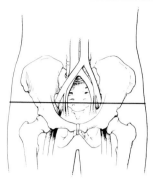

Body Coil
1, rectus abdominis
2, iliopsoas
3, sartorius
4, gluteus maximus
5, piriformis
6, colon
7, transverse abdominis

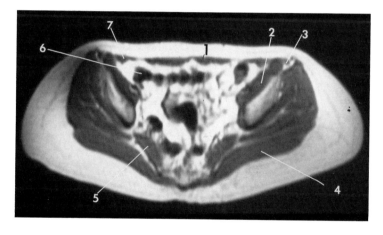

Body Coil (female)

1, rectus abdominis
2, iliopsoas
3, sartorius
4, tensor fascia lata
5, gluteus medius
6, gluteus minimus
7, gluteus maximus
8, piriformis
9, internal iliac branch vessels
10, uterus
11, colon
12, sciatic nerve
13, anterior inferior iliac spine
14, rectus femoris tendon
15, external iliac vessels

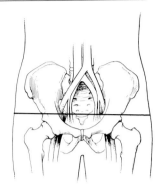

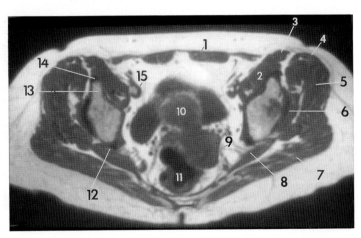

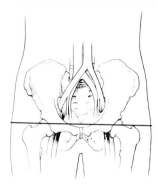

Body Coil

1, femoral artery and vein
2, iliopsoas
3, sartorius
4, tensor fascia lata
5, femoral nerve
6, gluteus medius
7, gluteus maximus
8, gemellus inferior
9, obturator internis
10, rectum
11, bladder
12, posterior acetabulum
13, greater trochanter
14, femoral head
15, tendon rectus femoris

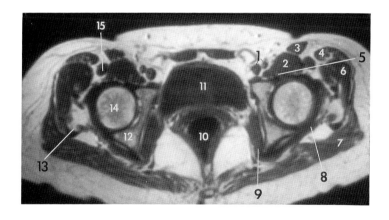

Body Coil
1, pubic symphysis
2, pectineus
3, femoral artery and vein
4, sartorius
5, tensor fascia lata
6, quadratus femoris
7, gluteus maximus
8, rectum
9, obturator internus
10, obturator externus
11, vastus lateralis
12, vastus intermedius
13, rectus femoris
14, iliopsoas

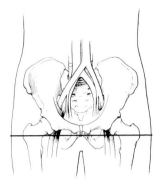

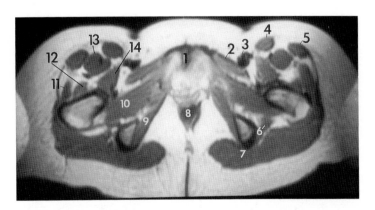

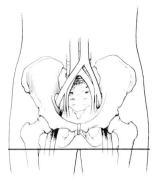

Body Coil (male)
1. gracilis
2. adductor longus
3. adductor brevis
4. adductor magnus
5. subtrochanteric portion of femur
6. gluteus maximus
7. origin hamstring muscles
8. vastus lateralis
9. vastus intermedius
10. rectus femoris
11. sartorius

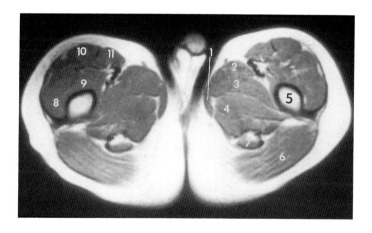

Body Coil (female)

1, L$_5$
2, uterus
3, rectum
4, vagina
5, retropubic fat
6, bladder
7, rectus abdominis

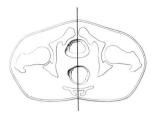

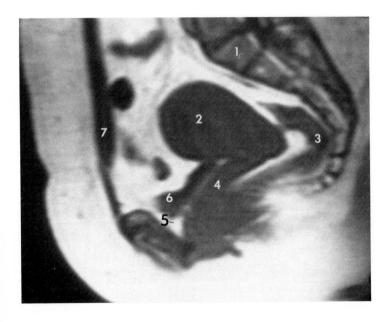

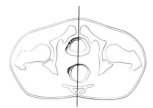

Body Coil (male)
1, rectum
2, levator ani
3, puborectalis
4, prostatic urethra and prostate
5, bulbo-spongiosis
6, urethra
7, testicle
8, retropubic fat
9, anterior inferior bladder wall
10, rectus abdominis
11, seminal vesicles
12, rectovesical pouch

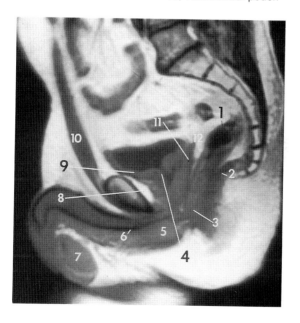

Body Coil

1, iliac vessels
2, gluteus maximus
3, obturator internus
4, obturator externus
5, adductor brevis
6, superior pubic ramus
7, bladder (lateral margin)
8, rectus abdominis

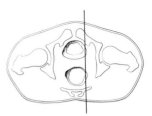

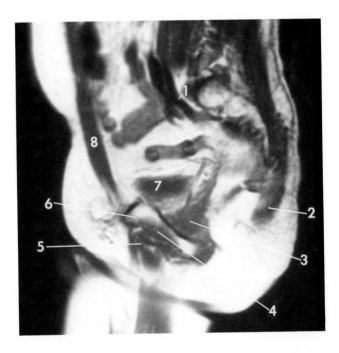

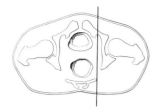

Body Coil

1, psoas major
2, erector spinae
3, internal iliac artery
4, piriformis
5, pudendal and vesical branches of internal iliac artery
6, gluteus maximus
7, obturator internus
8, inferior pubic ramus
9, quadratus femoris
10, adductor magnus
11, adductor brevis
12, adductor longus
13, obturator externus
14, superior pubic ramus
15, rectus abdominis

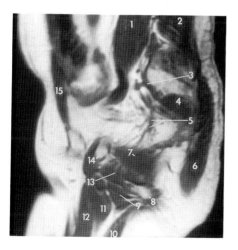

Body Coil
1, anterior superior iliac spine
2, iliacus muscle
3, tensor fascia lata
4, rectus femoris
5, sartorius
6, greater saphenous vein
7, femoral artery and vein
8, superior epigastric artery
9, pubic symphysis
10, lymph node

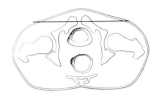

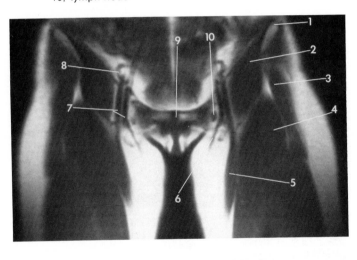

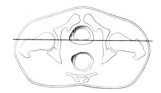

Body Coil

1, psoas muscle
2, iliacus muscle
3, acetabulum
4, obturator internus
5, obturator externus
6, adductor brevis
7, prostate
8, femoral head
9, gluteus minimus
10, gluteus medius

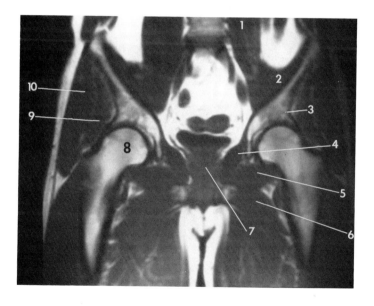

Body Coil
1, multifidus
2, longissimus
3, intercostalis and quadratus lumborum
4, L4 spinous process
5, left sacroiliac joint
6, gluteus minimus
7, piriformis
8, levator ani
9, rectum
10, sacral canal

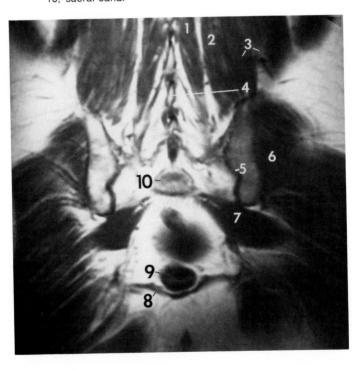

Body Coil
1, saphenous vein
2, adductor longus
3, gracilis
4, adductor brevis
5, adductor magnus
6, semitendinosus
7, gluteus maximus
8, vastus lateralis
9, vastus intermedius
10, vastus medialis
11, tensor fasciae latae
12, rectus femoris
13, deep femoral artery and vein and superficial femoral artery and vein
14, sartorius

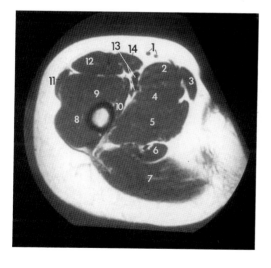

Body Coil

1, sartorius
2, superficial femoral artery and vein
3, saphenous vein
4, adductor longus
5, gracilis
6, deep femoral artery and vein
7, adductor brevis
8, adductor magnus
9, semitendinosus
10, fasciae latae
11, sciatic nerve
12, gluteus maximus
13, biceps femoris
14, vastus lateralis
15, vastus intermedius and medialis
16, rectus femoris

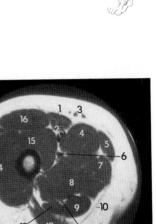

Body Coil
1, rectus femoris
2, vastus medialis
3, sartorius
4, greater saphenous vein
5, superficial femoral
   artery and vein
6, adductor longus
7, adductor magnus
8, gracilis
9, semimembranosus
10, semitendinosus
11, biceps femoris (long
   head)
12, sciatic nerve
13, biceps femoris (short
   head)
14, vastus lateralis
15, vastus intermedius

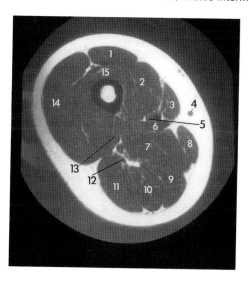

Body Coil
1, rectus femoris
2, vastus medialis
3, sartorius
4, greater saphenous vein
5, adductor longus
6, gracilis
7, adductor magnus
8, semitendinosus
9, biceps femoris (long head)
10, sciatic nerve
11, vein
12, biceps femoris (short head)
13, vastus lateralis
14, vastus intermedius

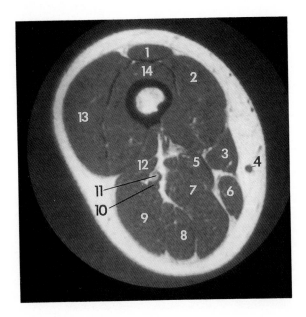

Body Coil
1, rectus femoris tendon
2, vastus medialis
3, femoral artery and vein
4, adductor magnus tendon
5, sartorius
6, greater saphenous vein
7, gracilis
8, semimembranosus
9, semitendinosus
10, sciatic nerve
11, biceps femoris
12, vein
13, vastus lateralis
14, vastus intermedius

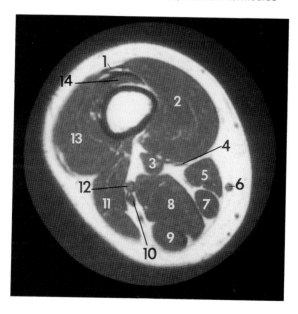

Body Coil
1, tendon quadriceps femoris
2, vastus medialis
3, sartorius
4, greater saphenous vein
5, gracilis
6, semimembranosus
7, semitendinosus
2, popliteal artery and vein
9, sciatic nerve
10, biceps femoris
11, vein
12, vastus lateralis
13, suprapatellar bursa

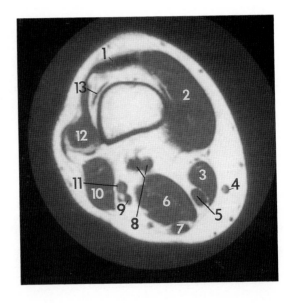

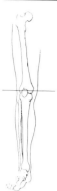

Surface Coil
1, patella
2, medial retinaculum
3, medial epicondyle
4, sartorius
5, greater saphenous vein
6, gracilis tendon
7, semitendinosus tendon
8, semimembranosus
9, gastrocnemius (medial)
10, popliteal artery and vein
11, tibial nerve
12, gastrocnemius (lateral)
13, biceps femoris
14, lateral epicondyle
15, vastus lateralis tendon

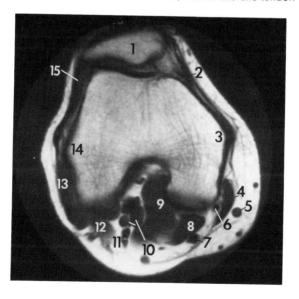

Surface Coil
1, infrapatellar fat
2, medial retinaculum
3, sartorius
4, gracilis tendon
5, greater saphenous vein
6, tendon semimembranosus
7, tendon semitendinosus
8, gastrocnemius (medial)
9, gastrocnemius (lateral)
10, common peroneal nerve
11, biceps femoris and tendon
12, anterior cruciate ligament

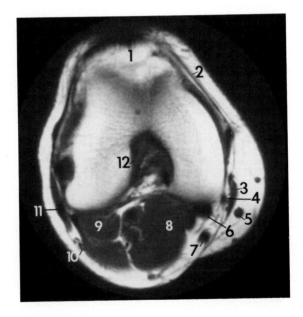

Surface Coil

1, patellar ligament
2, sartorius
3, gracilis tendon
4, semimembranosus tendon
5, semitendinosus
6, gastrocnemius (medial)
7, popliteal artery
8, tibial nerve
9, gastrocnemius (lateral)
10, common peroneal nerve
11, biceps femoris tendon
12, lateral collateral ligament
13, lateral meniscus

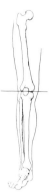

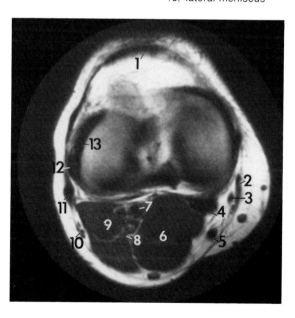

Surface Coil

1, patellar ligament
2, tibia
3, greater saphenous vein
4, semitendinosus tendon
5, gastrocnemius (medial)
6, small saphenous vein
7, gastrocnemius (lateral)
8, biceps femoris tendon

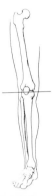

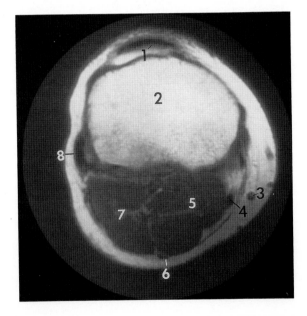

Surface Coil
1, medial gastrocnemius
2, medial femoral condyle
3, medial collateral ligament
4, medial meniscus
5, tibial spine
6, fibula
7, lateral meniscus
8, lateral femoral condyle

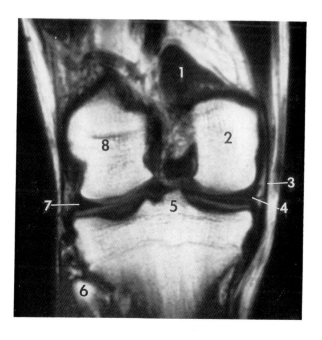

Surface Coil
1, medial gastrocnemius
2, articular cartilage of medial femoral condyle
3, posterior horn of medial meniscus
4, medial tibial plateau
5, anterior horn of medial meniscus

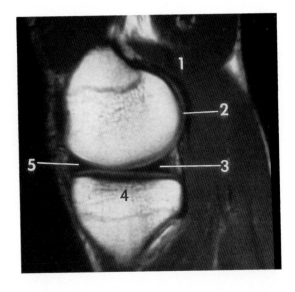

Surface Coil (A, neutral; B, 20° external rotation)

1, posterior cruciate ligament
2, popliteal artery
3, gastrocnemius
4, anterior cruciate ligament
5, tibial attachment of posterior cruciate ligament
6, patellar ligament

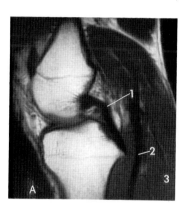 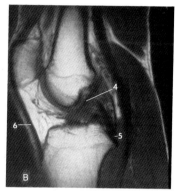

Surface Coil
1, lateral femoral condyle
2, posterior horn of lateral meniscus
3, fibular head
4, tibia
5, patellar ligament
6, anterior horn of lateral meniscus
7, infrapatellar fat
8, patella

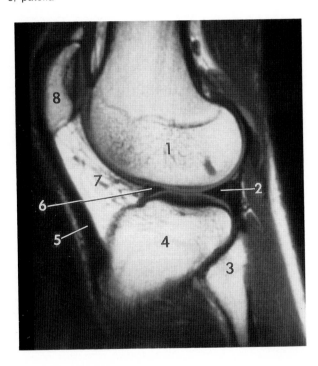

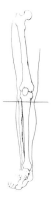

Surface Coil
1, tibial tuberosity
2, tibia
3, popliteus
4, greater saphenous vein
5, gastrocnemius (medial)
6, popliteal artery and vein
7, gastrocnemius (lateral)
8, soleus
9, fibula

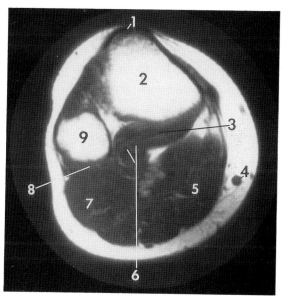

Surface Coil

1, tibia
2, tibialis posterior
3, posterior tibial artery and vein
4, gastrocnemius (medial)
5, soleus
6, gastrocnemius (lateral)
7, peroneal artery, vein, and nerve
8, peroneus longus and brevis
9, anterior tibial artery and vein and deep peroneal nerve
10, extensor digitorum longus and extensor hallucis
11, tibialis anterior

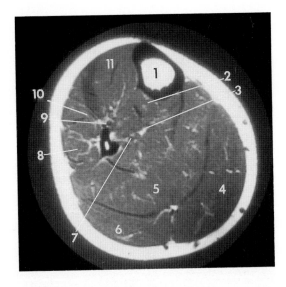

Surface Coil
1. tibia
2. tibialis posterior
3. flexor digitorum longus
4. gastrocnemius (medial)
5. soleus
6. gastrocnemius (lateral)
7. fibula
8. peroneus longus and brevis
9. extensor digitorum longus and extensor hallucis longus
10. tibialis anterior

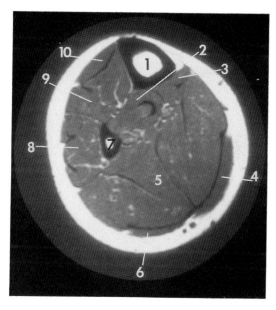

Surface Coil

1, tibia
2, tibialis posterior
3, flexor digitorum longus
4, greater saphenous vein
5, posterior tibial artery and vein and tibial nerve
6, gastrocnemius tendon
7, soleus
8, flexor hallucis longus
9, fibula
10, peroneus longus and brevis
11, extensor digitorum longus
12, extensor hallucis longus
13, tibialis anterior

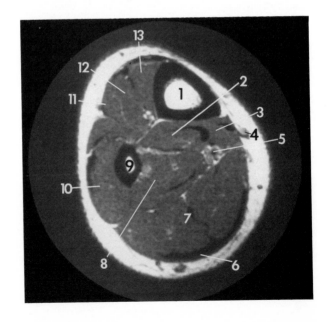

Surface Coil
1, tibialis anterior tendon and muscle
2, tibia
3, tibialis posterior
4, flexor digitorum longus
5, posterior tibial artery and vein
6, tibial nerve
7, soleus
8, gastrocnemius tendon
9, small saphenous vein
10, flexor hallucis longus
11, peroneus longus and brevis
12, fibula
13, extensor digitorum longus
14, extensor hallucis longus

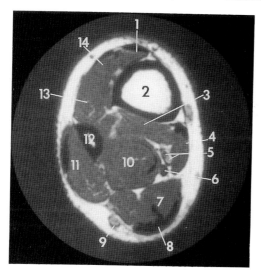

Surface Coil
1, tibialis anterior tendon
2, tibia
3, tibialis posterior tendon
4, flexor digitorum longus
5, posterior tibial artery and vein
6, tibial nerve
7, Achilles tendon
8, small saphenous vein
9, flexor hallucis longus
10, peroneus longus and brevis
11, fibula
12, extensor digitorum longus
13, extensor hallucis longus

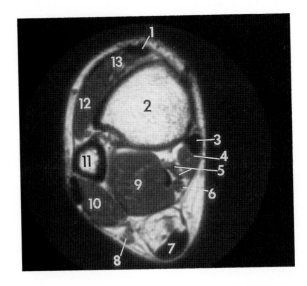

Surface Coil

1, tibialis anterior tendon
2, greater saphenous vein
3, medial malleolus
4, tibialis posterior tendon
5, flexor digitorum longus tendon
6, posterior tibial artery and vein
7, tibial nerve
8, flexor hallucis longus tendon
9, Achilles tendon
10, peroneus longus tendon
11, peroneus brevis tendon
12, fibula
13, extensor digitorum longus
14, extensor hallucis longus

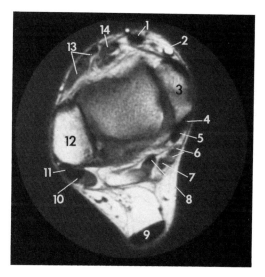

Surface Coil

1, extensor hallucis longus tendon
2, tibialis anterior tendon
3, talus
4, tibialis posterior tendon
5, sustentaculum tali
6, flexor digitorum longus tendon
7, flexor hallucis longus tendon
8, abductor hallucis
9, calcaneus
10, peroneus brevis and longus tendons
11, cuboid

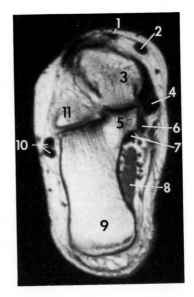

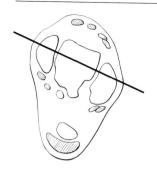

Surface Coil (mortise position)
1, tibial
2, medial malleolus
3, articular cartilage
4, talus
5, talocalcaneal interosseous ligament
6, calcaneous
7, lateral malleolus

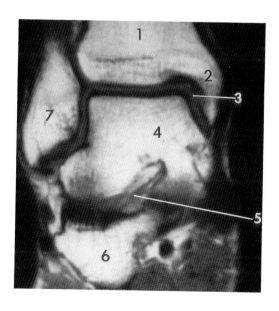

Surface Coil

1, medial malleolus
2, flexor digitorum longus tendon
3, tibialis posterior tendon

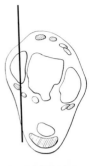

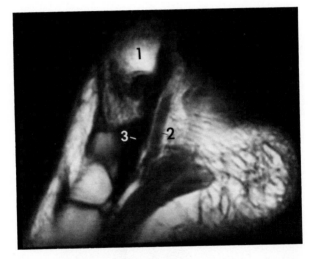

Surface Coil

1, tibia
2, sustentaculum tali
3, calcaneus
4, flexor hallucis longus tendon
5, navicular
6, talus

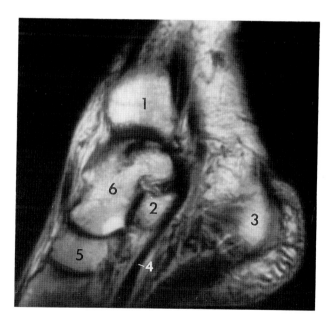

Surface Coil

1, tibia
2, flexor hallucis longus
3, pre-Achilles fat
4, Achilles tendon
5, calcaneus
6, plantar aponeurosis
7, flexor digitorum brevis
8, navicular
9, talus
10, tibialis anterior tendon

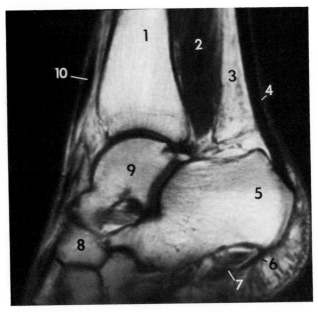

Surface Coil
1, fibula
2, peroneus longus tendon
3, peroneus brevis tendon

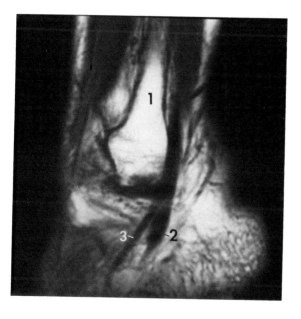

Bibliography

Auh YH, Rubenstein WA, Markesy JA, Zirinsky K, Whalen JA, Koyam E (1986): Intraperitoneal and paravesical spaces: CT delineation with US correlation. *Radiology* 159:311–317.

Auh YH, Rubenstein WA, Schneider M, Reckler JM, Whalen JA, Koyam E (1986): Extraperitoneal perivesical spaces: CT delineation with US correlation. *Radiology* 159:319–328.

Beltran J, Noto AM, Mosure JC, Weiss KL, Zuelzcr W, Christoforidis AJ (1986): The knee: surface coil MR imaging at 1.5 T. *Radiology* 159:747–751.

Cahill DR, Orland MJ (1984): *Atlas of Human Cross-Sectional Anatomy.* Lea and Febiger, Philadelphia.

Carter BL, Morehead J, Wolpert SM, Hammerschlag SB, Griffiths HJ, Kohn PC (1977): *Cross-Sectional Anatomy: Computed Tomography and Ultrasound Correlation.* Appleton-Century-Crofts, New York.

Grant JCB (1962): *Grant's Atlas of Anatomy.* Williams & Wilkins, Baltimore.

Holliday J, Saxon R, Lufkin RB, Rauschning W, Reicher M, Bassett L, Hanafel W, Barbaric Z, Sarti D, Glenn W (1985): Anatomic correlations of magnetic resonance images with cadaver cryosections. *Radiographics* 5(6):887–921.

Kieft GJ, Bloein JL, Obermann WR, Verbout AJ, Rozing PM, Doornvas J (1986): Normal shoulder: MR imaging. *Radiology* 159:741–745.

Wagner M, Lawson TL (1982): *Segmental Anatomy Applications to Clinical Medicine.* MacMillan, New York.